Smithsonian

In Gear

Amy Pastan and Linda McKnight

Smithsonian

Gear

Collins

An Imprint of HarperCollinsPublishers

SMITHSONIAN IN GEAR. Copyright © 2007 by Amy Pastan and McKnight Design, LLC.

HarperCollins books may be purchased for educational, business, or sales promotional use. For information please write: Special Markets Department, HarperCollins Publishers, 10 East 53rd Street, New York, NY 10022.

FIRST EDITION

Designed by Linda McKnight, McKnight Design, LLC

Edited by Anita Schwartz

The authors would like to thank Ellen Nanney of Smithsonian Business Ventures for coordinating this project. Her efforts made this book series possible.

Library of Congress Cataloging-in-Publication Data

Pastan, Amy.
   Smithsonian in gear / Amy Pastan and Linda McKnight.
      p. cm.
   ISBN 978-0-06-125150-4
   1. Gearing--Exhibitions. I. McKnight, Linda. II. Title.

TJ184.P36 2007
629.04—dc22   2007018456

07 08 09 10 11   TP   10 9 8 7 6 5 4 3 2 1

The authors would like to thank the following Smithsonian museums, research centers, and offices for their assistance and cooperation in the making of Spotlight Smithsonian books:

Anacostia Community Museum
Archives of American Art
Arthur M. Sackler Gallery
Cooper-Hewitt, National Design Museum
Freer Gallery of Art
Hirshhorn Museum and Sculpture Garden
National Air and Space Museum
National Air and Space Museum's
   Steven F. Udvar-Hazy Center
National Anthropology Archives
National Museum of African Art
National Museum of American History,
   Kenneth E. Behring Center
National Museum of the American Indian
National Museum of Natural History
National Portrait Gallery
National Postal Museum
National Zoological Park
Smithsonian American Art Museum
   and its Renwick Gallery
Smithsonian Institution Libraries
Smithsonian Institution Archives
Smithsonian Astrophysical Observatory
Smithsonian Center for Folklife
   and Cultural Heritage
Smithsonian Environmental Research Center
Smithsonian Photographic Services
Smithsonian Tropical Research Institute
Smithsonian Women's Committee,
   Office of Development

Smithsonian in Gear

Where can you see everything from a 1920s passenger locomotive to a nineteenth-century mail wagon, Eli Whitney's cotton gin to a Model T Ford Roadster, a Singer sewing machine to the Space Shuttle *Enterprise*? From hot rods to low riders, cool bikes to sleek rockets, and groundbreaking inventions to quirky gadgets—the Smithsonian has it all. Among the wealth of items in its collections are many treasures for those who love things that "go." But with seventeen museums and a zoo in Washington, D.C., two museums in New York City, various research centers, and more than 137 million objects in all, the Smithsonian can be overwhelming—even for those who have the opportunity to come often. Now *Smithsonian in Gear* offers a selection of unique items from the collections and allows you to see them as no ordinary visitor can. In these pages you can experience highlights from the exhibits as well as see lesser-known materials that are seldom on view. Jump from a car canvas by Stuart Davis at the Smithsonian American Art Museum to the transportation collection at the National Museum of American History. Find a historic photograph of the first rocket launch at Cape Canaveral from the National Air and Space Museum next to a digital image of a Pakistani painted truck from a recent Smithsonian Folklife Festival. Enjoy the world's largest museum, cultural, and scientific complex without planning a trip, fighting the crowds, or paying for gas. And return as often as you like.

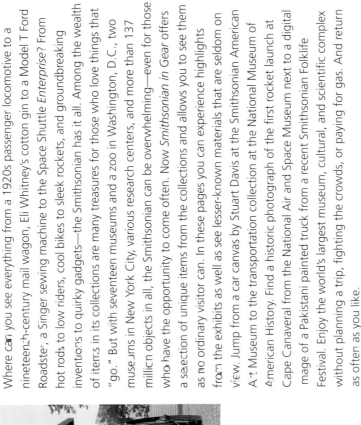

Hart Parr Tractor #3

1903
National Museum of American History
Division of Work and Industry

Photograph for Jones and Laughlin Steel Corp. Advertising Campaign

1965
Arthur D'Arazien Industrial Photographs
Ektachrome print
Archives Center, National Museum of American History

This engaging parade of people, riding, carrying, or wheeling everyday items made of steel (a baby carriage, hoe, pogo stick, watering can, lawnmower), is an advertisement for the Jones and Laughlin Steel Corporation. But aside from promoting steel products, the ad also makes a statement. The smokestacks that dot the horizon are symbols of American industry and confirm the notion of America's economic strength.

Silk Road Painted Truck

Smithsonian Folklife Festival 2002
Color photograph
Smithsonian Center for Folklife and
Cultural Heritage

The ancient Silk Road—a vast trade
route—crossed the mountains and
deserts of Central Asia, connecting
East Asia and the Mediterranean. In
2002, the Smithsonian Folklife Festival
celebrated the arts and cultures of the
people of the Silk Road lands. Once,
the roads in Afghanistan and Pakistan
were alive with brightly colored wagons
and carts. This elaborately painted
truck from Pakistan comes out of that
tradition. Poems, sayings, landscapes,
and intricate designs cover its surfaces,
making it a magnet for curious visitors.

Transportation

c. 1951
Saul Steinberg
Pen and ink, watercolor, and collage of pho-
tomechanical reproductions on paper
Hirshhorn Museum and Sculpture Garden
Gift of Joseph H. Hirshhorn, 1966
©2007 Saul Steinberg Foundation/Artists
Rights Society (ARS), New York

Saul Steinberg (1914-1999) includes
many forms of transportation in this
witty drawing executed in various
media—pen and ink, watercolor,
and photo reproduction. There are
hovering planes and a smoking ocean
liner at top, a train careening across
the horizon, a bicycle parked in what
appears to be a station, and solitary
travelers, with and without baggage,
dotting the landscape. The trunk in the
foreground—on a wheeled trolley—
seems to have survived many journeys.

Jim Beam – Caboose

1986

Jeff Koons

Stainless steel and bourbon

Hirshhorn Museum and Sculpture Garden

Fractional and Promised Gift of Paul A. and
Anastasia Polydoran, Des Moines, Iowa,
1996

© Jeff Koons

Artist Jeff Koons (b. 1955) developed
a fascination with a series of glass
decanters that bourbon producer Jim
Beam manufactured in the 1980s. The
decanters were designed as train cars,
from locomotives to cabooses. Koons
had copies of the decanters fabricated
in steel. This Jim Beam caboose is both
a captivating visual object and a symbol
of American consumerism.

Ice-Cream Vendor

After 1870
Antoin Sevruguin
Glass plate negative
Freer Gallery of Art and
Arthur M. Sackler Gallery Archives
Myron Bement Smith Collection
Gift of Katharine Dennis Smith

A wheeled ice cream cart located
somewhere in the Near East is the
focus of this fascinating photograph
by Antoin Sevruguin (1830s-1933).
It is part of the Myron Bement Smith
Collection at the Freer and Sackler
Galleries, which houses almost 75,000
images. Sevruguin was born at the
Russian Embassy in Tehran, where his
father was a diplomat. His artful images
capture nineteenth-century Iran—both
the royal court of the shahs, as well as
the lives of common people. He sold his
work to local customers as well as to
foreign visitors who bought the pictures
as souvenirs.

Evans Dual-Purpose Streamlined Auto-Railers

1930s
Detroit, Michigan
Smithsonian Institution Libraries

If any place could be called "in gear," it would be Detroit, Michigan, original home of the U.S. auto industry. This futuristic "auto-railer" was created in Detroit to help railroads compete for passenger and freight traffic. Designed to go on rail or road, the "auto-railer" never made it from drawing board to assembly line, but this image from the Smithsonian Libraries reveals a sleek art deco style that conveys a sense of progress.

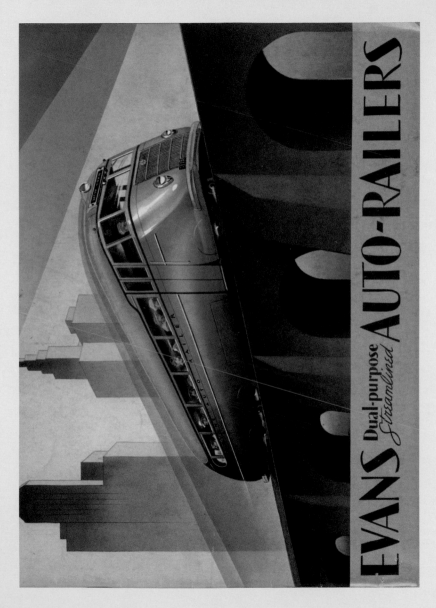

Hendey Machine Tools Catalog Cover

c. 1936
Torrington, Connecticut
National Museum of American History
Division of Work and Industry

This loose cover of the Hendey Machine Tools catalog dates from the 1930s. Clearly handled numerous times, it was once attached to the full catalog, which was used for ordering Hendey products. It is now at the Smithsonian, part of a huge collection of similar trade literature that researchers use. This particular piece was preserved for its illustrative value, allowing researchers to see the lathes, planers, and other types of equipment that were available in an earlier century.

Studebaker Ad

1953
Smithsonian Institution Libraries

Brothers Henry and Clement Studebaker opened the Studebaker blacksmith shop in South Bend, Indiana, in 1852. By 1857, they were making carriages as well, and during the Civil War, they furnished supply wagons to the U.S. Army. Studebaker's products were highly prized by average consumers as well as the country's elite. Abraham Lincoln rode to Ford's Theatre in a Studebaker carriage the evening he was assassinated. The company successfully made the transition to automobile production in the early twentieth century. In the 1950s, Studebaker cars were on the cutting edge of design, with their bullet-nose front ends and wrap-around rear windows.

The new American car with the European look

**THE NEW
1953 STUDEBAKER**

Travelling Carnival, Santa Fe

1924
John Sloan
Oil on canvas
Smithsonian American Art Museum
Gift of Mrs. Cyrus McCormick

John Sloan and his wife visited Santa
Fe for the first time in 1919; they
would return every year for the next
thirty years. Sloan may have witnessed
this scene several times, as traveling
carnivals came through town on a
regular basis, with the carousel and
Ferris wheel being the main attractions.
Here the illuminated carousel against
the dark sky creates an eerie effect,
but the woman in pink doesn't seem
to notice. She rides her horse with
abandon as a crowd of cowboys, young
girls, and older Spanish women in
shawls look on.

Ferris Wheel

1893
Hubert Howe Bancroft
from *The Book of the Fair*
Chicago and San Francisco
Smithsonian Institution Libraries
World's Fair Collection
Gift of Larry Zinn

The World's Columbian Exposition of 1893 in Chicago offered a view of the world's standing in science, art, and industry as the nineteenth century was closing and the twentieth century was dawning. Progress in all fields was spectacular as the beautiful images in this rare volume convey. This color illustration shows grand pavilions of all designs dotting the landscape, which is dominated by one of the Exposition's highlights—a huge Ferris wheel rising above the fair's midway.

Freight Cart

After 1860

Kusakabe Kimbei

Hand-colored photographic print

Freer Gallery of Art and
Arthur M. Sackler Gallery Archives
Henry and Nancy Rosin Collection of Early
Photography of Japan
Purchase and gift of Henry and Nancy Rosin

Kusakabe Kimbei (1841–1934) operated a studio in Yokohama from the early 1880s until 1918. He learned his art as an apprentice to Baron Raimond von Stillfried, an Austrian who worked in Japan from 1812 to 1889. This hand-colored albumen print by Kimbei from the Freer and Sackler archives shows Japanese workers laboring over a heavy cart. Kimbei produced scenes such as this for tourists who considered them examples of genuine Japanese culture. His beautiful stylized views of Japanese life continue to be popular.

Leavitt Canceling Machine
Patent Model

1879
Metal, wood
United States
National Postal Museum

The Leavitt canceling machine reduced the manual labor involved in postmarking and canceling mail pieces. Feed rollers introduced "faced mail" into the machine one piece at a time. Each piece progressed to a rotating canceling die hub and ring die. Once the postmark and cancellation impressions were made, the pieces passed to a stacker, where they were retrieved by a machine operator. While the Leavitt machine could only accept mail pieces of a standard size, it pioneered the technology of mail cancellation, spurring other inventors to make further improvements.

Acme Voting Machine

1880
Bridgewater, Connecticut
National Museum of American History
Division of Politics and Reform

Intended to ensure an honest vote, the Acme voting machine was an improvement over the open-slot box. It has a tabulator, activated by a lever mechanism that releases the ballot into the box. This machine was manufactured in Bridgewater, Connecticut, about 1880.

Two-slot Postage Vending Machine

1950s–1960s
New York
Metal (steel) and paint
National Postal Museum

Back in the days when postage stamps cost only four or five cents, this machine was regularly in use. Designed to sit on a countertop, it was a simple device—insert your money, turn the crank, and take your stamps from the slot. The slots at bottom served as coin returns. The National Postal Museum has several postage vending machines in its collection.

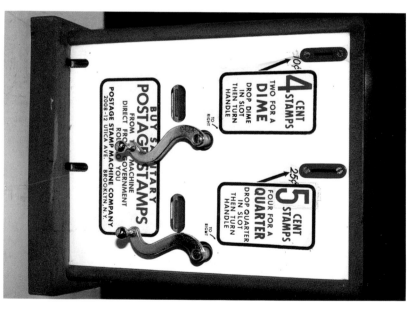

Jackpot Machine

1962
Wayne Thiebaud
Oil on canvas
Smithsonian American Art Museum
Museum purchase made possible by the
American Art Forum and gift
of an anonymous donor
Art ©Wayne Thiebaud/Licensed by VAGA,
New York, NY

The machine of hope—the jackpot—boldly confronts the viewer, almost daring him or her to put a coin in the slot and pull the lever. Painted in bold red, white, and blue—the colors of the American flag and the American dream—the painting lures those looking to win. But the promise of a prize is uncertain—only two of the machine's three tokens line up.

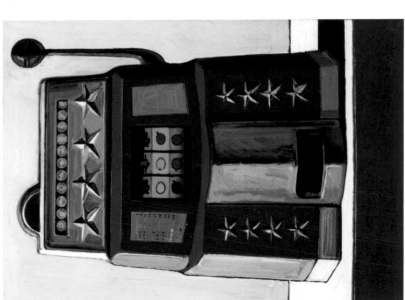

Steamboat "Chesapeake"

c. 1887
Unidentified photographer
Black and white photographic print
National Postal Museum

In the early 1800s riverboats like the
one pictured here helped carry mail to
the southern and western United States.
This paddle wheel steamboat is at an
unidentified port. While such boats
were fast, and often the most reliable
transport for mail, poor weather,
accidents, and delays were frequent
impediments to maintaining regular
schedules.

Burden's Wheel

c. 1890s
Unidentified photographer
Albumen print
National Museum of American History
Division of Work and Industry

Henry Burden, a Scot who emigrated to the United States in 1819, was a brilliant inventor. Some years after settling in Troy, New York, and becoming superintendent of the Troy Iron and Nail Factory, Burden devised machines to manufacture railroad spikes and horseshoes, replacing work that had been done by hand. Under his direction, the factory became very profitable and he soon became its sole owner. But Burden is perhaps best remembered for his extraordinary water wheel. Built to generate power for his horseshoe factory, it was 62 feet in diameter and produced up to 482 horsepower.

Ford Model AA 1-ton Parcel Post Truck

c. 1931-1954
Wood, metal, glass, paint
United States
National Postal Museum

When this 1931 Ford Model AA truck was purchased, drab olive green was the signature color of the Post Office Department. The trucks carried letter carriers to their daily routes or transported mail between post offices and railway stations. During the Great Depression and World War II, the Post Office Department did not purchase many new vehicles, so several trucks were on the road for longer than expected—some even stayed in service until the 1950s.

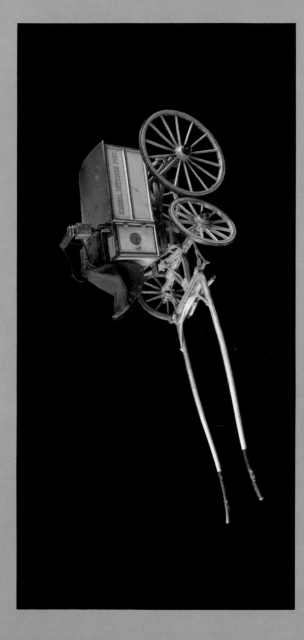

German Mail Wagon Model

Late 19th century
Wood, leather, metal
National Postal Museum

In late nineteenth-century Germany, the mail may have been delivered on a wheeled yellow wagon. This model of a four-wheeled Kaiser Deutsche Post cart was created by Robert Liebzcher of Dresden, Germany. Such a wagon would have been pulled by a single horse. German mail vehicles and mailboxes continue to be painted this distinctive shade of yellow.

Mechanical Toy

c. 1862

Tin

George W. Brown & Co., Connecticut

National Museum of American History

Division of Home and Community Life

An extensive collection of antique toys may be found at the National Museum of American History. Subjects include all kinds of objects on wheels—horse-drawn vehicles, mail and fire trucks, police wagons, and farm equipment. This charming example belonged to a little girl in Washington, D.C., during the Civil War. When wound with a key, the wheels turn and the carriage goes on its way. Manufacturer George W. Brown & Co. of Connecticut produced these popular wind-up toys in the mid- to-late nineteenth century.

Sewing Machine
Patent Model

1854
Patented by Isaac Singer
National Museum of American History
Division of Home and Community Life

Isaac Singer and New York lawyer Edward Clark established I.M. Singer & Co. in 1851. The ready-made clothing industry quickly embraced Singer's lockstitch sewing machine and soon "Singers" of various types were found in factories and sweatshops. Singer later offered payment plans, allowing home seamstresses to buy affordable machines as well. With models designed to suit different needs and tastes, the company manufactured two million machines by 1877 and expanded worldwide. The National Museum of American History has both patent models and early machines by Singer.

Steam Tricycle in Front of North Entrance to Smithsonian Institution Building

1888
Unknown photographer
Black and white photographic print
Smithsonian Institution Archives

Today's tourists at the Smithsonian Castle may arrive there on anything from rollerblades to skateboards, but these visitors in the 1880s had their own interesting means of conveyance—a steam tricycle. With fringed sunshade, this odd-looking vehicle used a steam boiler to propel the chassis of a tricycle. Various models existed—some coal-fired and others fueled by petroleum. Eventually, the invention of the internal combustion engine led to the replacement of steam.

World War I Tank outside National Museum

c. 1920s
Unknown photographer
Black and white photographic print
Smithsonian Institution Archives

In the 1920s, when the structure now called "Arts and Industries" was "The United States National Museum (USNM)," a park outside the building displayed various artifacts too large to install in exhibition halls. This impressive World War I "baby tank" was such an item. Other objects from the Smithsonian war collection could be seen in the same setting. Today, lovely gardens surround most of the Arts and Industries building.

Installation of Locomotive in National Museum of History and Technology

1961
Unknown photographer
Black and white photographic print
Smithsonian Institution Archives

Many who have actually seen the "1401," a 280-ton Pacific-type passenger steam locomotive, at the Smithsonian's National Museum of American History, have wondered—how did they get such an enormous train into the building? As shown in this photograph, the locomotive was installed while the building was still under construction. Workers attached it to an elaborate pulley system so that it could be towed into the transportation hall. The impressive train was built in 1926 by the Richmond Virginia Works of the American Locomotive Company.

Sightseeing Bus

1906
Unknown photographer
Black and white photographic print
Smithsonian Institution Archives

As the nation's capital, Washington, D.C., has always been a popular city for tourists. The Smithsonian is a large part of the attraction. Today, tour buses ferry millions of people to and from the National Mall to visit the museums each year. Apparently, crowds were also common back in 1906, when this sight-seeing vehicle set out on a route from G Street, NW. Complete with lap robes to keep patrons warm, the service also provided guides equipped with megaphones to narrate the tour.

Pickering Wind Tricycle

c.1900-1909
Unknown photographer
Gelatin silver print
National Air and Space Museum Archives
William J. Hammer Scientific Collection

Among the estimated 17 million
photographs in the National Air and
Space Museum's archives is this quirky
image. An unidentified man rides a
contraption known as the "Pickering
wind tricycle," an invention that left
no discernable mark on history. In fact,
we know virtually nothing about this
innovation, except what we can glean
from this early photograph.

Edmond Poillot

c. 1910–1912
Unknown photographer
Gelatin silver print
National Air and Space Museum Archives

Edmond Poillot has a friendly companion at the helm of this precarious-looking aircraft. Poillot was born in 1888. The photo was taken several years before his death in 1914. He and his faithful friend are piloting a Voisin biplane at Farman's Aviation School in Mourmelon, France.

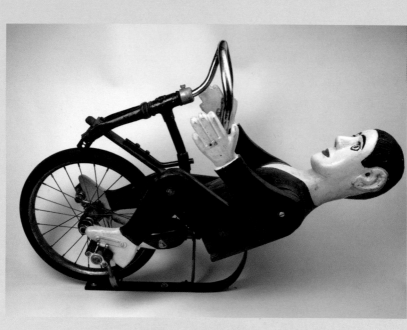

Bicycle Shop Sign

Early 1930s
Louis Simon
Carved and painted wood, gesso, metal and
rubber bicycle parts, marbles, and metal
hardware
Smithsonian American Art Museum
Gift of Herbert Waide Hemphill, Jr. and
museum purchase made possible by Ralph
Cross Johnson

This bicycle shop sign was made by
champion motorcycle racer and folk
artist Louis Simon (1884-1970). Simon
opened a shop in 1922 that specialized
in bike sales and repair. This sign
announced his business to the public.
Installed on the exterior wall of the shop,
it had a motor to make both the wheel
turn and the legs move up and down.

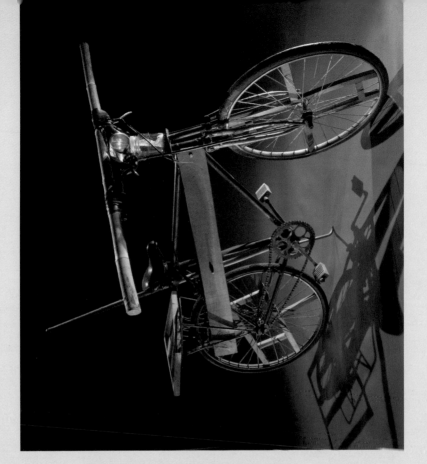

Bicycle

c. 1960s
Vietnam
Metal, wood, rubber
National Museum of American History
Division of Military History and Diplomacy

This bicycle was used by Viet Cong soldiers on the Ho Chi Minh trail during the Vietnam War. The Viet Cong was a guerrilla force that fought against the United States and South Vietnam during the conflict and was supported by the North Vietnamese Regular army. They often used makeshift weapons, had a variety of uniforms, and avoided traditional combat, making them an elusive enemy.

Hand-Drawn Hose Reel

c.1855
Robert Frazier
National Museum of American History
Division of Home and Community Life

This ornamented hose reel dates from about 1855 and could carry 400 to 600 feet of leather hose. Firefighters adorned such equipment with elaborately painted panels, gilt ornaments, and etched, colored lamps. Pieces such as this one were used primarily in parades. The inscription 1838 probably refers to the date the hose company was organized. The reel belonged to the Reliance Hose Company No. 3 of Bethlehem, Pennsylvania. The hose reel is part of the CIGNA Firefighting and Maritime Collection at the National Museum of American History. CIGNA began as the Insurance Company of North America in 1792. In the early nineteenth century, the company began collecting, commissioning, and preserving maritime and firefighting objects related to its various lines of business. The artifacts date from around 1750 to the mid-twentieth century.

Indian Motorcycle with Princess Sidecar

1923

George Hendee and Carl Oscar Hedstrom
National Museum of American History
Division of Work and Industry

Bicycle racer George Hendee and engineer Carl Oscar Hedstrom founded a company with the idea of building the first "motor-driven bicycle for everyday use." Indian motorcycles were produced from 1901 to 1953, with various models tailored to gentlemen riders, police officers, and thrill seekers.

Schwinn Panther

1953

National Museum of American History
Division of Work and Industry
Courtesy of Schwinn Bicycles
Schwinn® is used by permission of Pacific Cycle, Inc.

Kids loved the Schwinn Panther, particularly when it had fun handlebar accessories like the ones pictured here. The Schwinn Bicycle Company was founded by Ignaz Schwinn in Chicago in 1895 and became the dominant manufacturer of American bicycles through the twentieth century. Schwinn bikes had superb craftsmanship, sleek styling, and best of all, were affordable for consumers—mostly children—who loved chrome fenders, balloon tires, and a large saddle.

Men and Machines, from the Black Biker Series

1983
Dan Williams
Color coupler print on paper
Smithsonian American Art Museum
Museum purchase
© 1983 Dan Williams

Now these are machines! One wonders how many gears they actually have. Taken by photographer Dan Williams in 1983, the images from this photo series show African American bikers in impressive motorcycle gear at a festival in Louisville, Kentucky. The prints form part of the contemporary photography collection at the Smithsonian American Art Museum.

Industrial Life

1941
Seymour Fogel
Tempera on paper
Smithsonian American Art Museum
Transfer from the General Services
Administration

Seymour Fogel experimented with many styles throughout his career, from realism to abstract expressionism. He produced numerous murals for the Works Progress Administration (WPA) and the Treasury Section of Painting and Sculpture from 1934 to 1941. This study shows men at work charting the course of the future—the gears at center symbolize the wheels of progress. Fogel learned mural art from a master, serving as an apprentice to Diego Rivera on Rivera's famous Rockefeller Center commission.

Automotive Industry (mural, Detroit Public Library)

1940
Marvin Beerbohm
Oil on canvas mounted on board
Smithsonian American Art Museum
Transfer from the Detroit Public Library

Marvin Beerbohm originally painted this work for the technology building of the Detroit Public Library. In planning the composition, he not only consulted numerous books about the automotive industry from the library's collection, but also visited auto plants in the Detroit area to achieve a composite view of the entire production process. Beerbohm's picture reveals an environment in which man is in harmony with machines.

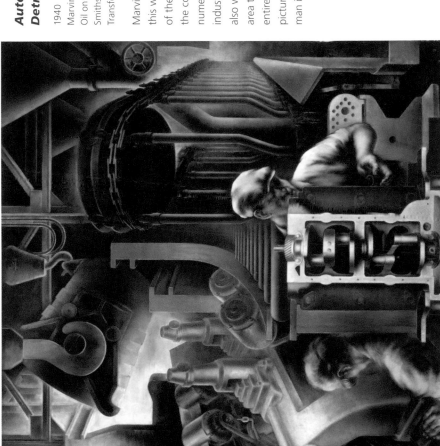

Paper Workers

1934
Douglass Crockwell
Oil on canvas
Smithsonian American Art Museum
Transfer from the U.S. Department of Labor

Painted in 1934 as the nation struggled through the Great Depression, this powerful work pays homage to the efforts of American workers to keep the country running. A huge paper milling machine dominates the canvas and dwarfs the four workers who smooth the sheet as it is wound onto a giant roller. The workers are machinelike as well, and seem intrinsically linked to the production process. Using subtle tones and abstract forms, the artist conveys America's strong devotion to modernity and progress.

Eli Terry Mass-Produced Box Clock

c. 1816
Plymouth, Connecticut
National Museum of American History
Division of Work and Industry

Early in the nineteenth century, Eli Terry and his associates Seth Thomas and Silas Hoadley saw a need for low-cost domestic clocks. They applied water-powered machine technology to clock-making and turned out factory clocks with wood movements instead of expensive and scarce brass ones. Terry designed this box clock to sit on a mantel or shelf. It became enormously popular, transforming the economy of western Connecticut (where he worked) from agricultural to industrial, with more than one hundred firms soon making machine-made clocks.

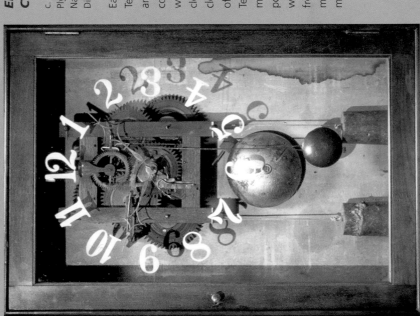

Model of Eli Whitney's Cotton Gin

1782
National Museum of American History
Division of Home and Community Life

This is the model for Eli Whitney's famous cotton gin, which was patented in 1793. The cotton gin is a mechanical device that removes seeds from the cotton plant, a process that revolutionized the cotton industry in the American South. Whitney's initial hand-cranked gin could produce up to fifty pounds of cleaned cotton daily, which boosted the economy of the Southern states and, some say, further entrenched them in the system of slavery, which provided their labor force.

Samuel Slater's Spinning Frame

1790
National Museum of American History
Division of Home and Community Life

British-born Samuel Slater was apprenticed to a textile factory master in England and acquired many trade secrets. Defying England's laws, which prohibited emigration of engineers, he traveled to the United States and helped found the first successful water-powered textile mill in America. Slater also built this 48-spindle spinning machine to spin cotton. It was first operated by him on December 20, 1790, at Pawtucket, Rhode Island, and is the oldest piece of textile machinery in America.

Pennsylvania Railroad
Locomotive at the Altoona
Repair Facility

c. 1868
Unidentified artist
Albumen print
Smithsonian American Art Museum
Museum purchase from the Charles
Isaacs Collection made possible in part by
the Luisita L. and Franz H. Denghausen
Endowment

This albumen print is part of the Charles
Isaacs Collection of photographs at the
Smithsonian American Art Museum. The
collection documents the first century
of photography from 1839—the year
the medium was invented—to 1939.
Railroads played a major role in the
growth and development of the United
States during those years. This grand
locomotive dominates the image, and
the viewer can almost feel its weight,
power, and speed.

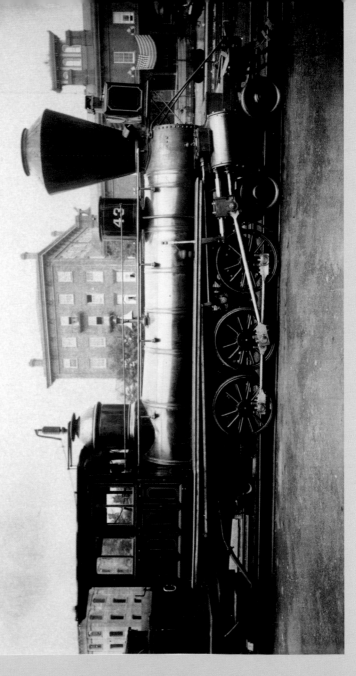

Huey Helicopter 091
(Bell UH-1 Iroquois)

1965

Manufactured by Bell Helicopter Company
National Museum of American History
Division of Military History and Diplomacy

Four thousand Bell H-1 helicopters, nicknamed "Hueys," were deployed in Vietnam during the Vietnam War. This particular Huey served in 1966 with the 173rd Assault Helicopter Company, known as the "Robin Hoods." The helicopter continued to fly after the war, until 1995. In 2002, 091 was the centerpiece of a documentary film about Vietnam veterans and their stories.

Electric Streetcar, Capital Traction, Co., No. 303

Made 1898, used 1898-1913
Manufactured by the American Car
Company, St. Louis, Missouri
National Museum of American History
Division of Work and Industry

This bright yellow, electric-powered streetcar was one of several that provided regular service for passengers along the 7th-Street Line in Washington, D.C., for more than a decade. The *No. 303* trundled on rails running up 7th Street, between Arsenal Street at the Potomac River wharves, to Boundary Street, now Florida Avenue. Because overhead wires were banned in the District of Columbia, an underground conduit supplied the power. The *No. 303* has a extension "shoe" on its underside that connected to the conduit through a slot between the rails. Similar streetcars ran on other routes in the city.

Ford Model T Roadster

1926

Manufactured by the Ford Motor Company
National Museum of American History
Division of Work and Industry

The Ford Motor Company sold more than 15 million Model Ts between 1908 and 1927, partially due to the car's affordability. Customers could choose from various styles, including this sporty roadster. Ford perfected mass production techniques. Using a moving assembly line, he produced cars at affordable prices, giving him an edge over his competitors. When General Motors introduced the concept of the annual model change—which encouraged buyers to regularly turn in their old vehicles for new ones—Ford's sales dropped. The company stopped producing the Model T in 1927.

Garage No. 1

1917
Stuart Davis
Oil on canvas
Hirshhorn Museum and Sculpture Garden
Gift of Knoll International, 1980
Art © Estate of Stuart Davis/Licensed by
VAGA, New York, NY

Stuart Davis (1892-1964) generally found his subjects in everyday scenes. Although he experimented with cubism, he later executed pure abstract images. This early work shows Davis' interest in the American landscape. The old car and quaint gas pump, though painted as little more than geometric shapes, clearly evoke a nostalgia for a simpler time.

Ford Country Squire Station Wagon

1955
Manufactured by the Ford Motor Company
National Museum of American History
Division of Work and Industry

In the 1950s this station wagon became the car of the suburbs and the middle class—similar to minivans and SUVs today. Nancy and George Harder, a couple from southern California, drove their five kids to school in this car. It was the vehicle used for picnics and beach outings and to transport the family dog. When the Harder kids were teenagers, they learned to drive this classic, which became a symbol of family activity and togetherness. It now evokes nostalgic memories from Smithsonian visitors.

Pit Crew Ballet

1984
Jeff Tinsley
Color photograph
National Museum of American History
Division of Work and Industry

The extraordinary choreography in this
photograph by Jeff Tinsley is performed
by racer Richard Petty's crew as they
complete a four-tire change during a pit
stop at the Talladega 500. Petty drove
this 1984 Pontiac stock car to his 200th
victory on the Grand National stock car
circuit at Daytona. Fans are delighted to
see the car at the Smithsonian's National
Museum of American History.

"Dave's Dream," Lowrider

1978-1992
1969 Ford LTD
National Museum of American History
Division of Work and Industry

In 1978 David Jaramillo of New Mexico bought a 1969 Ford LTD and began converting it into "Dave's Dream"—a lowrider that he hoped would be a prizewinner at auto shows. Later that year, Jaramillo died, and his family decided to continue work on "Dave's Dream" as a memorial to him. From 1979 to 1982, the family entered the car in many shows, often taking "Best Lowrider" award. Jaramillo had installed a new V8 engine, added the sunroof, and started painting the body. After his death, family members finished some of the remaining modifications. It has a red velour interior, iridescent exterior paint, and a hydraulic suspension system to make it "hop."

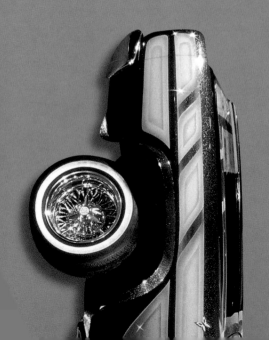

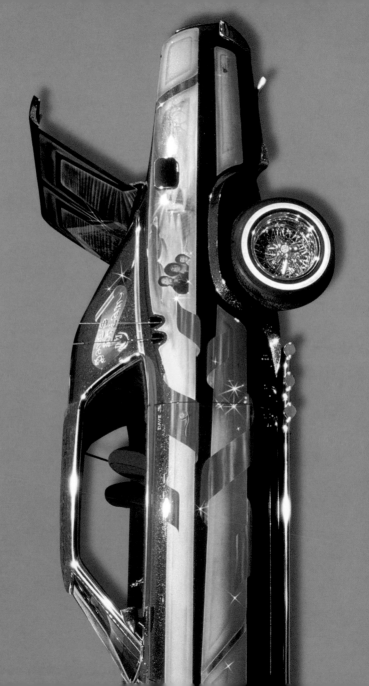

Crowds Cheer Henry Farman III Biplane

c. 1911
C.H. Detrich
Toned gelatin silver print
National Air and Space Museum Archives

In the early days of aviation, air shows were popular events. In this early photo crowds cheer in the grandstand as a Henry Farman III biplane soars overhead. The plane may have been piloted by Louis Poulhan at an air show in the New York metropolitan area in late May 1911.

First Launch from Cape Canaveral, Florida

July 24, 1950
U.S. Air Force
Gelatin silver print
National Air and Space Museum Archives

The Bumper 8 rocket was launched at Cape Canaveral, Florida, on July 24, 1950. It comprised a German V-2 missile lower stage and WAC-Corporal upper stage. Cameramen assembled in the foreground, recording the historic spectacle for future generations.

Shuttle Orbiter Enterprise
Off-loaded from 747

1983
Courtesy of NASA
National Air and Space Museum

Visitors to the National Air and Space Museum's Steven F. Udvar-Hazy Center can see the first Space Shuttle *Enterprise*. This was a test vehicle designed to operate in the Earth's atmosphere—not space. It was flown atop a Boeing 747 shuttle carrier and released for piloted free-flights and landings to check out its performance. These tests were necessary preludes to the Space Shuttle *Columbia* launch—the first orbital flight by a space shuttle—in 1981. After completing its work as a test vehicle, NASA briefly put *Enterprise* on display at a Paris Air Show and World's Fair and then delivered it to the Smithsonian in 1985.

SR-71 **Blackbird Landing**

1990
Color photograph
National Air and Space Museum

The SR-71 *Blackbird* is seen with its chute deployed, touching down on the runway of Dulles International Airport. It is now in the collection of the Smithsonian's National Air and Space Museum and displayed at the Steven F. Udvar-Hazy Center. The SR-71 *Blackbird* is the fastest aircraft propelled by air-breathing engines. Its performance and operational achievements placed it at the forefront of aviation technology developments during the Cold War. Called "Blackbird" because of its black surface, the plane's paint was formulated to absorb radar energy, radiate heat generated by aircraft skin friction at high speeds, and conceal the aircraft at high altitudes and at night.

Credits

Smithsonian image numbers are followed by names of the photographers, when known.